A KEY-GUIDE TO MAMMAL SKULLS AND LOWER JAWS

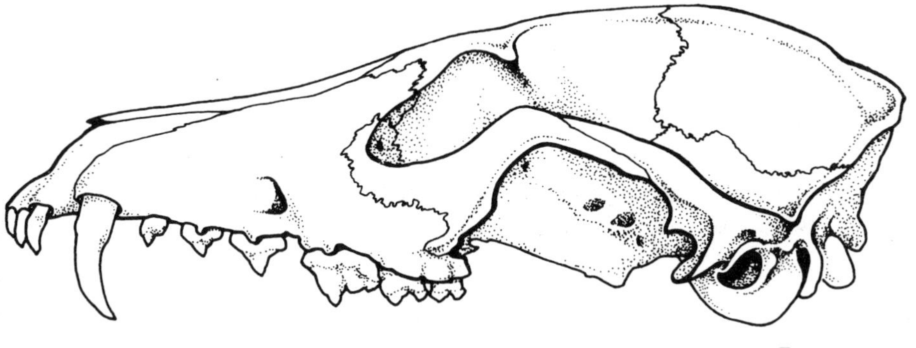

A KEY-GUIDE TO MAMMAL SKULLS AND LOWER JAWS

by Aryan I. Roest

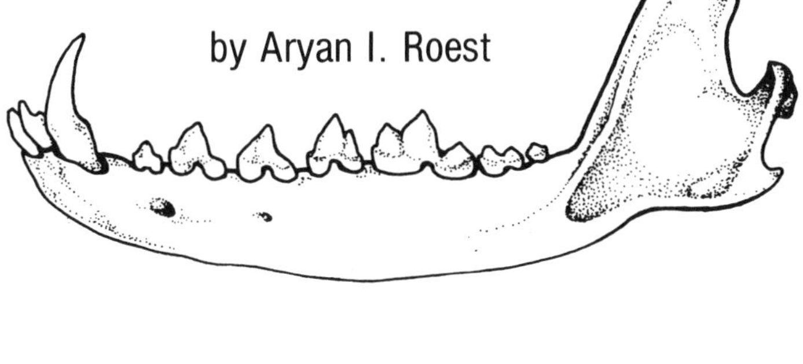

A nontechnical introduction for beginners

Author's note:

Should any user of this booklet think of ways in which it might be improved, I would greatly appreciate a letter indicating what such improvements might be. Thank you.

A. Roest

The author and publisher express their thanks to Ms. Marguerite Costigan for the cover and title page illustrations.

Published by: Mad River Press, Inc.
Rt. 2, Box 151 B
Eureka, CA 95501

Printed by: Eureka Printing Co., Inc.
106 T Street
Eureka, CA 95501

ISBN 0-916422-71-2

TABLE of CONTENTS

LIST of ILLUSTRATIONS

PREFACE

Anyone who has spent time tramping through fields, hiking in the mountains, or walking along beaches has occasionally found the remains of an animal long since dead. Bones and bits of fur are all that is left of what was once a living mammal. The first question in the viewer's mind is usually "What was it?" This booklet is designed to help identify the remains and answer that question. It may also be useful to those who have jobs which bring them into regular contact with mammal remains, such as wildlife biologists and game wardens. It can also be helpful to teachers and students of natural history, who are often called upon to give an identification for a skull which has been brought to them.

This booklet is <u>not</u> intended for specialists or experts who seek detailed information on skull features of particular species, or who require exact identifications for small mammal skulls. In most cases such experts may already know the identity of the usual 'pick-up' skulls that they examine. Rather, it is intended for outdoor-oriented people who wish to learn how to recognize mammal skulls at a glance. Although most of the mammals whose skulls can be identified with these keys are wild species, many of the more common domestic species can also be identified with them.

These keys will help identify the skulls of most wild and domestic mammals which occur in the United States and southern Canada. The skulls of many mammals which occur in northern Canada or in northern Mexico can also be identified, but the skulls of some of the species occurring in those regions will not key out. Skulls of most species occurring elsewhere in the world cannot be identified with these keys.

In preparing this booklet I have had the pleasure of handling a great many nicely cleaned skulls from a wide variety of mammals. Such skulls are attractive and can tell a great deal about the mammals from which they came. I have also had the benefit of the fine keys prepared by other mammalogists for the proper identification of various species of mammals. Some of these are listed in the references at the back of this booklet. Finally, I have had the good fortune to expose a great many of my students to earlier versions of these keys. As they ran into problems, the keys would be modified or improved, so now I trust that they work fairly well. To all these -- skulls, other mammalogists, and students -- I extend my sincere thanks.

Aryan I. Roest

Biological Sciences Department
California Polytechnic State University
San Luis Obispo, CA 93407

HOW TO USE THE KEYS

In preparing the keys in this booklet, a minimum of technical terms has been used. Nevertheless, some terms are used and should be learned. All terms are listed and explained in the Glossary, and many are also illustrated in the drawings. It is a good idea to become familiar with the more frequently used terms before trying to use the keys. To do this, read through the Glossary and study the labelled drawings for a few minutes.

Size is often used as a distinguishing feature. Most measurements are only approximations. They are given in millimeters, but in most cases the equivalent in inches is also provided. Keep in mind that 25 mm equals 1 inch, 50 mm is about 2 inches, and so on.

All measurements are taken with the skull or jaw in a position such that the cheek teeth are in a nearly horizontal line. Note the drawings on pages 8 and 22. A scale is provided on the back cover of this booklet, but a small, flexible plastic ruler is easier to use. The skulls of young mammals may not be large enough to provide measurements so they will key out correctly.

Actual identification of a mammal skull or jaw is made with the help of a "key". A key is a selection of features arranged in a sequence of alternatives, from which a choice must be made at each step in the sequence. By following through the series of choices correctly, you are guided to an identification. The keys in this booklet are arranged as a series of numbered steps, and at each step two or more contrasting features are described, each indicated by _a_, _b_, _c_, or _d_. At most steps only two choices are offered.

For example, assume you have found a skull and wish to key it out to an identification. In the SKULL KEY (page 9), step 1 offers a choice among several sizes involving the length of the skull: 1-a is a large skull, 1-b is for medium skulls, etc. If the skull you have is about 3 inches long (75 mm), your choice among these selections would be 1-b (for skulls 3-6 inches long). This choice is followed by the instruction to _go to step 19_. At step 19 (on page 14), you must again choose, between skulls with or without canine teeth. Your skull has canines, so you choose 19-a; you must now count the number of incisor teeth on each side of the skull midline (there are three in your skull), and then proceed to step 20. Your skull has its postorbital processes distinctly in front of the skull midpoint (20-b), so you move on to step 25, on page 15. Here you must count the number of cheek teeth; your skull has 6, so you have reached an identification --- your skull is that of a RACCOON.

As indicated above, teeth are counted on one side only. If the key choice specifies '4 cheek teeth', this means 4 teeth on one side of the skull or jaw, although the total number of cheek teeth is obviously 8. Tooth counts in the skull (upper jaw) may differ from those in the lower jaw.

HOW TO USE THE KEYS

Often skulls or jaws found in the field are broken, and teeth may be missing. In such cases, remember that skull are symmetrical, and a broken part may still be intact on the other side. Missing teeth are indicated by holes (alveoli) in the supporting bone, but while some teeth have a single root, marked by one hole, the cheek teeth may have as many as 5 roots, marked by as many holes. It is not always easy to get a proper count of the number of teeth in such cases, but it is usually possible to determine the number by careful examination. Some mammals have very tiny teeth; these are sometimes missed in counting, but again, careful study should reveal their presence.

Broken skulls or jaws are often difficult to identify, since important features may be missing. In hoofed mammals, the front bones (premaxillas) of the skull are often broken and lost; this affects identification right away in terms of the size of the skull. In such cases, try to determine whether a bone _is_ broken and not there, and allow for it in making measurements.

If both skull and lower jaw are available, run them both through the appropriate keys. They should both key to the same species, and thus provide a check on your identification. Some of the smaller rodents are particularly difficult to identify, and such forms should be regarded as only _tentatively_ identified until they can be checked with more formal keys or other sources of identification. See the REFERENCES listed in the back.

The scientific names of the mammals which can be identified in these keys are given in a list of names at the back of the booklet. For many of the smaller mammals, several species may be included within a single common name. WOODRAT, for example, is a kind of mammal which actually includes 9 different species which occur in the United States. This key will not help you determine to which of these species a skull may belong. An identification can be narrowed down by considering woodrat species distributions. Maps which show the distribution of most species are available in many books on mammals, particularly in field guides. Examination of the maps showing woodrat distribution in A FIELD GUIDE TO THE MAMMALS, by Burt and Grossenheider (see REFERENCES) will help you in your identification.

To determine which species of Woodrat (for example) may be the source of a skull which keys out to WOODRAT, the distribution maps show that only 4 species occur in California (assuming the skull was found in that state): the Desert Woodrat, White-throated Woodrat, Bushy-tailed Woodrat, and Dusky-footed Woodrat. If the skull came from north of San Francisco Bay, the only possible species it could be is the Dusky-footed Woodrat.

GLOSSARY of TERMS

Many of the features described here in the Glossary are also illustrated in the drawings of skulls and jaws presented on many of the following pages.

Alveolus	- a hole or socket into which the root of a tooth fits.
Angle	- the lower back corner of the lower jaw.
Anterior	- toward the front of the skull or lower jaw.
Antler base	- flat-topped bump on the frontal bone which, in male deer, supports the antler. Also called the <u>pedicel</u>.
Braincase	- that part of the skull, behind the orbits, which contains the brain.
Bulbous	- bubble-like, rounded convexly.
Bulla	- bulbous projection of the earbone.
Canine	- one of the four teeth in the front corners of the mouth. Canine teeth are usually large and pointed in meat-eating species, but may be small, or even absent, in plant eaters. They are the first (the most anterior) teeth in the maxilla, or upper jaw bone. In humans, they are sometimes called 'eye-teeth' because they are located just below the eyes.
Cheek teeth	- teeth rooted in the maxilla, behind the canines, along the sides of the mouth. They include both premolars and molars, and may have from 1 to 5 roots, depending on their size. In the lower jaw <u>all</u> teeth behind the canines (or incisors, if canines are absent) are also cheek teeth.
Concave	- curving inward, away from the surface; forming a depressed area, or dip, or sometimes a pit.
Condyle	- rounded projection with a connecting surface on a bone which can meet a similar surface on another bone to form a moveable joint. The <u>condylar process</u> of the lower jaw is where it joins the skull.
Convex	- curving outward, away from the surface; swollen or bulging.
Coronoid process	- part of the lower jaw which rises up from the back end of the horizontal ramus, from near the last cheek teeth in the ramus.

GLOSSARY

Cranium	- the skull without the lower jaws.
Depth of ramus	- vertical distance from the upper surface of the ramus of the lower jaw to the lower surface, or edge.
Earbone	- structure on the lower surface of the skull which contains the middle ear. It may be bulbous or relatively flat, and is sometimes called the tympanic bulla. The passage from the outer ear (the ear canal or external auditory meatus) leads to it from the side of the skull.
Flange	- extra plate or development which projects beyond the main surface of the bone involved.
Foramen	- hole or opening in a bone through which nerves and blood vessels may pass.
Foramen magnum	- large opening at the back of the skull, in the occipital bone, through which the spinal cord leaves the brain. The occipital condyles lie on either side of it, to form the joint where the skull meets the first bone of the backbone.
Forehead	- region between and above the orbits of the eyes, primarily involving the frontal bones.
Frontal bones	- form the top, front portion of the braincase, where they lie behind the nasal bones, between and behind the orbits, and in front of the parietal bones.
Horn core	- porous bone projecting from the frontal bone which, in the living animal, is covered by the horn proper.
Incisor	- tooth in the front of the mouth, between the canines. Incisors of the upper jaw grow from the premaxilla; all other upper teeth grow from the maxilla.
Infraorbital foramen	- a large foramen, or hole, in the side of the maxilla, usually in front of, or below, the orbit.
Interorbital width	- Distance between the upper edges of the orbits, measured across the top of the skull.
Interparietal bone	- an unpaired bone lying between the two parietal bones and the occipital bone; useful in noting the difference between hares and rabbits.
Inward	- toward the center, or midline, of the skull.

GLOSSARY

Jaw - the lower jaw, or mandible; it is formed of the two <u>dentary</u> bones, one on each side.

Lacrimal bone - forms the front edge of the orbit, between the maxilla and the nasals and frontals. It is large and distinct in hoofed mammals, but small and inconspicuous in most other species.

Lacrimal pit - a depression or dimple in the lacrimal bone.

Lateral - toward the side, away from the skull midline.

Length of jaw - greatest length of the lower jaw, measured from the front of the incisors to the most posterior point of the jaw, either at the angle of the jaw or at the condyle (or condylar process).

Length of skull - greatest length of the skull, measured from the front of the incisors, or from the most anterior part of the premaxilla, or nasals (whichever is most anterior), to the most posterior part of the skull at the occipital condyles, occipital crest, or sagittal crest.

Midpoint of skull - about halfway between the front of the premaxilla and the back of the occipital crest.

Maxilla - bone forming major portion of the upper jaw on each side, and supporting the upper canines and the cheek teeth. Joins the premaxillas to the front, the nasals at the top, and provides the base for the front end of the zygomatic arch.

Nasal bones - form the upper surface of the rostrum, or muzzle, lying above the nasal opening. They are located in front of the frontal bones, and above and between the premaxillas and maxillas.

Occipital bone - large, broad bone forming the back of the skull, the back of the braincase; contains the large foramen magnum near its lower edge.

Occipital condyle - small knob, with distinct articulating surface, on either side of the foramen magnum, where the occipital bone joins the first vertebra of the backbone. There are two occipital condyles in mammals; birds and reptiles have only one.

Occipital crest - ridge formed where the parietal bones join the occipital bone, across the top, back part of the braincase. Also termed <u>lambdoidal crest</u>.

GLOSSARY

Orbit - the eye socket, bounded on the front by the maxilla, on the upper side by the frontal, and on the lower edge by the zygomatic arch. The size of the orbit can be determined by measuring its diameter, from rim to rim. The length of the orbit is its diameter from front to rear, horizontally. The height of the orbit is its diameter vertically, from top to bottom edges.

Outward - toward the side, laterally, away from the midline of the skull.

Palate - surface forming the roof of the mouth, behind the incisors and between the canines and cheek teeth.

Paroccipital processes - distinct downward projections of bone, between the earbones and the occipital condyles.

Parietal bones - form the top, back portion of the braincase, behind the frontal bones and in front of the occipital bone.

Posterior - toward the back, or rear, of the skull or lower jaw.

Postorbital process - lateral projection of the frontal bone that marks the back, upper edge of the orbit. There may be a postorbital process projecting upward on the upper surface of the zygomatic arch also, which marks the back, lower edge of the orbit.

Premaxilla - forms the most anterior part of the skull, the front tip of the upper jaw, in front of the maxilla. Supports the incisor teeth, and curves upwards toward the nasal bones.

Ramus - the main horizontal portion of the lower jaw, which supports the lower cheek teeth.

Root - the part of a tooth which is imbedded in the bone which supports it. Some teeth have only one root, some may have as many as five. Each root fits into a socket, or hole (the alveolus).

Rostrum - that part of the skull in front of the orbits, or in front of the braincase; also referred to as the muzzle.

Sagittal crest - lengthwise ridge on the top of the braincase, at the suture where the two parietals join. Usually runs anteriorly from the midpoint of the occipital crest.

GLOSSARY

Suture - the visible joint, or line, where two bones meet firmly, so no movement is possible. Sutures mark the edges of the various bones of the skull.

Width of skull - maximum distance across the zygomatic arches; also termed the breadth of the skull, or the zygomatic width or breadth.

Zygomatic arch - curved bones forming an arch along the side of the skull below the orbit, extending from the maxilla to a point near the external opening of the ear canal; sometimes referred to as the cheekbone, containing the jugal bone.

RACCOON

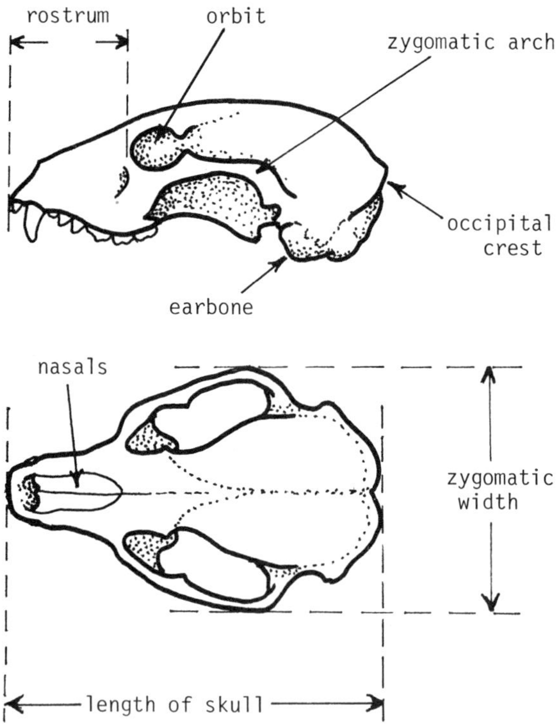

The drawings in this guide are not drawn to the same scale; they do indicate the shape and special features of skulls and lower jaws.

KEY-GUIDE to SKULLS

1 - a. Large skull, over 150 mm (6") long: go to step ----------- 2
b. Medium skull, 75-150 mm (3-6") long: go to step ---------- 19
c. Small skull, 25-75 mm (1-3") long: go to step ------------ 32
d. Tiny skull, less than 25 mm (1") long: go to step -------- 47

2 - a. Orbit (eye socket) closed at back by a bony bar formed of fused postorbital processes; no canines, <u>OR</u> canines about same size, or smaller than, cheek teeth: ----------------------------- 3
b. Orbit open at the back; canines large, prominent: --------- 10

3 - a. Skull over 300 mm (1 foot) long: ------------------------- 4
b. Skull less than 300 mm long: ----------------------------- 7

4 - a. Incisors present; infraorbital foramen above 2nd or 3rd cheek teeth:
- Orbit smoothly oval in shape; horizontal diameter of orbit about 12 mm (½") greater than vertical diameter: ------------ HORSE
-- Orbit flat along back edge; horizontal and vertical diameters about equal: ------- DONKEY
b. NO incisors; infraorbital foramen above 1st cheek tooth: -- 5

<u>HORSE</u>

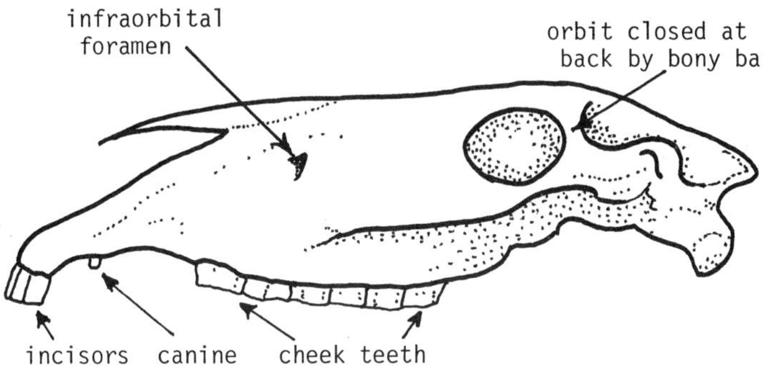

<u>COW</u>

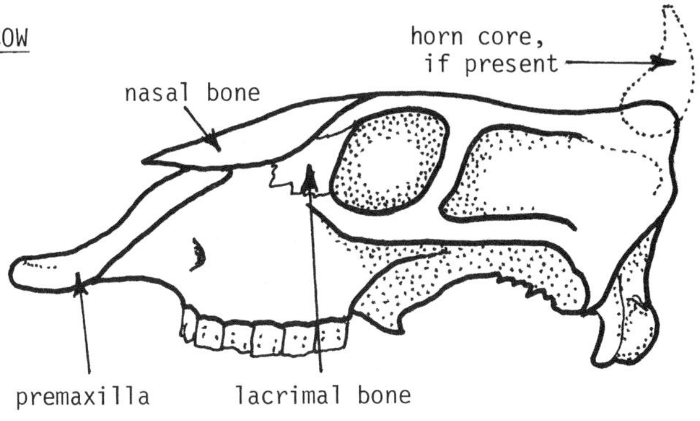

5 - a. Lacrimal bone separated from nasal bone by a wide space, exposing underlying bony membranes: -------- 6

b. Lacrimal bone contacts nasal, OR is separated from it by only a narrow slit; underlying membranous bone not exposed:

- Upper end of premaxilla just reaches nasal bone: ------------------------ DOMESTIC COW

-- Upper end of premaxilla does not reach or approach nasal bone: -------------------- BISON

--- Upper end of premaxilla has long zone of contact with nasal bone: ------------------ 7

6 - a. Nasal bones short, barely reaching a point above the 1st cheek tooth; skull over 400 mm (16") long: --- MOOSE

b. Nasal bones project forward beyond the 1st cheek tooth; skull less than 400 mm long; low, rounded canine teeth may be present:

- Premaxilla about 12 mm (½") wide; back edge of nasal bones form a point near the midline: --------------------- ELK

-- Premaxilla less than 10 mm wide; back edge of nasal bones form a straight line across skull -------------- CARIBOU

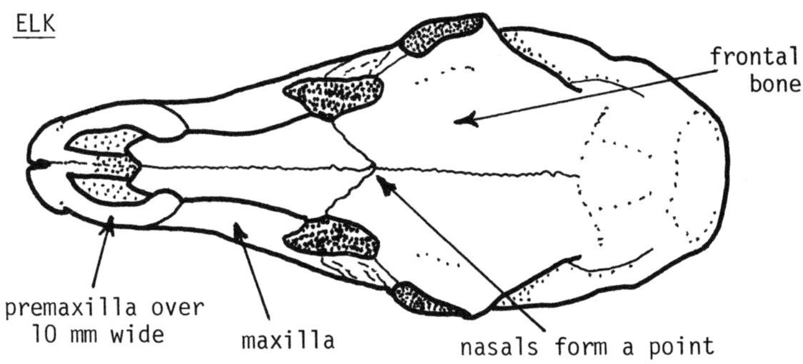

7 - a. Horn core, or bump, directly above the orbit; large foramen on inner base of horn core: ------- PRONGHORN

b. NO horn core, OR horn core (or antler base) behind the orbit; NO single large foramen on inner base of horn core, although there may be several smaller ones present: ------------------ 8

8 - a. Lacrimal bone contacts nasal bone, or is separated from it by a space of less than 10 mm (½"): ------------ 9

b. Lacrimal bone separated from nasal bone by a wide space of more than 10 mm, exposing the thin, membranous bones lying underneath: ------------------- DEER

- Lacrimal bone with a pit or depression in it which is more than 10 mm (½") deep: --------------------------------- MULE DEER

-- Lacrimal pit shallow, less than 10 mm deep: ------------------------- WHITE-TAILED DEER

<u>DEER</u>

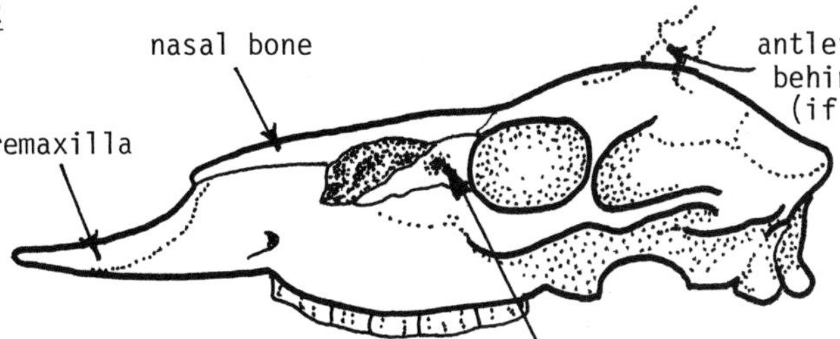

9 - a. Upper end of premaxilla contacts nasal bone; frontal and nasal bones have a straight or slightly concave profile; horn core (if present) with a ridge, or edge, along the inner front surface: ----------- DOMESTIC GOAT

b. Upper end of premaxilla does <u>NOT</u> reach nasal bone, or does so just barely; nasal bones somewhat arched, or convex (roman-nosed):

- NO horn core, <u>OR</u> horn core with a cross-section like a rounded triangle; lacrimal bone has a pit or depression in it: - WILD or DOMESTIC SHEEP

-- Horn core slender, erect, rounded or oval in cross-section; <u>no</u> pit or depression in lacrimal: ---- MOUNTAIN GOAT

10 - a. Orbit <u>behind</u> midpoint of skull; a bony flange at the base of the canines: ---------------------------- 11

b. Orbit <u>at</u> or <u>in front of</u> skull midpoint; <u>NO</u> extra bone growth at base of canines: ------------------ 12

11 - a. 3 upper incisors; canines curve outward to the sides; occipital crest extends back about as far as the occipital condyles; braincase flat across the top, between pinched-in sides; paroccipital processes project downward more than 25 mm (1"); most sutures visible, distinct: ------ WILD or DOMESTIC PIG

(Domestic pigs have a more concave, scooped profile along the nasal region, in contrast to a rather straight profile in wild types.

b. 2 upper incisors; canines point straight downward; occipital crest projects back beyond the level of the occipital condyles; braincase rounded across the top, in front of pinched-in sides; paroccipital processes short, less than 15 mm long; distinct grooves on rostrum; no sign of sutures in most adult skulls: --------------- PECCARY

12 - a. Nasal bones short to medium length, less than 5x as long as they are wide: --------------------------- 13

b. Nasal bones very long and narrow, more than 5x as long as they are wide:

- Skull less than 200 mm (8") long; palate <u>more</u> than 2x as long as wide; forehead flattish: ----------------------- COYOTE

-- Skull more than 200 mm long; palate <u>less</u> than 2x as long as wide; forehead somewhat swollen, rounded, over orbits: ----- WOLF

--- Some skulls, usually less than 200 mm long, with palates less than 2x as long as wide, and with rounded, convex foreheads: --- DOMESTIC DOG

(some dog breeds have skulls which will <u>not</u> key out here, or anywhere)

<u>COYOTE</u>

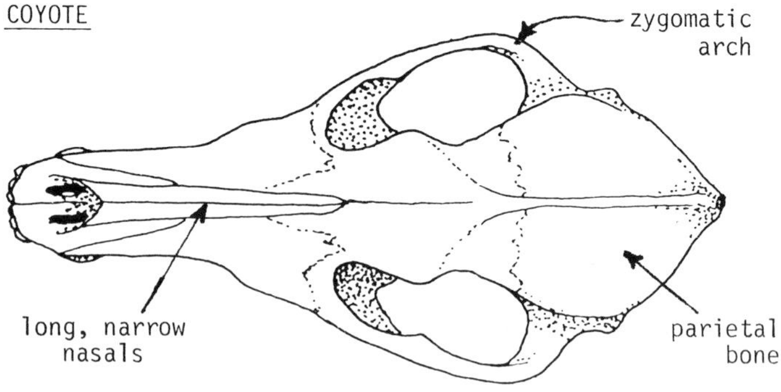

SKULLS

13 - a. Cheek teeth vary in size and shape; last or next-to-last cheek teeth larger than those toward front of mouth; occipital crest is usually most posterior part of skull: --- 14

b. Cheek teeth all essentially similar in size and shape; occipital condyles are most posterior part of skull: ------------------------------ 15

14 - a. Rostrum short, about as long as the diameter of the orbit; 2 (of 4 total) cheek teeth quite large, with sharp points and cutting edges toward front and back: --------------- MOUNTAIN LION

b. Rostrum about 2x as long as diameter of orbit; last 2 or 3 (of 5-6 total) cheek teeth have low, rounded bumps or cusps: Bears ---

- Last and next-to-last cheek teeth are less than 30 mm and 20 mm long, respectively: -------------------------- BLACK BEAR

-- Last and next-to-last cheek teeth are more than 31 mm and 21 mm long, respectively: -------------- BROWN, or GRIZZLY BEAR

--- If occipital condyles project back about as far as the back edge of the occipital crest, and if the last and next-to-last cheek teeth are 26 and 20 mm long respectively: ---- POLAR BEAR

15 - a. Postorbital processes distinct; cheek teeth often stained brown: ----------------------------------- 16

b. NO postorbital process, or a postorbital process is only suggested by a small bump; cheek teeth white, not stained brown: ----------------- 17

16 - a. Last cheek tooth separated from next-to-last by a space of about 10 mm (½") or more: --- NORTHERN SEA LION

b. Last cheek tooth NOT separated distinctly from next-to-last - space between them is less than 5 mm or so: ---

- Nasals 3x longer than wide; front of premaxilla, viewed in profile, slopes up toward the rear: ---- CALIFORNIA SEA LION

-- Nasals less than 2½x as long as wide; front of premaxilla, seen in profile, rises almost vertically: --- NORTHERN FUR SEAL

17 - a. 3 upper incisors; upper end of premaxilla just reaches nasal bone: ------------------------------------ 18

b. 2 upper incisors; upper end of premaxilla does NOT reach, or even come close to the nasal bone; skull usually over 200 mm (8") long; found along the Pacific coast: -------- ELEPHANT SEAL

SKULLS

18 - a. Each cheek tooth with 2-3 distinct points; skull usually less than 200 mm (8") long: ---- HARBOR SEAL

b. Cheek teeth peg-like, with single rounded points; skull usually over 200 mm long: ------- GRAY SEAL

19 - a. Canines large, prominent; NO toothless space in front of the cheek teeth: ---

- With 3 incisors on each side: ------------------ 20

-- With 5 incisors on each side: ------------- OPOSSUM

b. NO canines; large toothless space between incisors and cheek teeth: ----------------------------- 28

c. NO canines, NO incisors; 7-8 blunt, peg-like cheek teeth: ----------------------------------- ARMADILLO

20 - a. Postorbital process at, or near, skull midpoint: --------- 21

b. Postorbital process distinctly in front of skull midpoint: 25

21 - a. Rostrum short, less than the diameter of the orbit; 3-4 cheek teeth; palate wider than length of cheek tooth row: Cats -------------------------------- 22

b. Rostrum longer than diameter of orbit; 6-7 cheek teeth; palate narrower than length of cheek tooth row: -- 23

22 - a. Skull over 100 mm (4") long; NO tiny tooth between canine and 1st large cheek tooth: ---

- Interorbital width less than 30 mm: -------- BOBCAT

-- Interorbital width more than 30 mm: ---------- LYNX

b. Skull less than 100 mm long; a tiny peg-like tooth behind the canine: ------------------- DOMESTIC CAT

23 - a. Forehead essentially flat; canines slender, oval in cross section; 4th cheek tooth 2x longer than it is wide: Foxes --------------------- 24

b. Forehead rounded, convex, roundness extending onto back of nasals; canines with sharp edges on front and back; 4th cheek tooth about as long as it is wide: ---------------------- COATI

24 - a. Prominent ridges extend from postorbital processes back nearly to the occipital crest; ridges on both sides form a "U" or lyre shape on top of the skull: -------------- GRAY FOX

b. Low ridges extend back from postorbital processes, forming a "V" shape on top of the skull: ---------------------------------- RED FOX

(If the skull is less than 120 mm long, it is probably from either a SWIFT FOX or a KIT FOX)

SKULLS

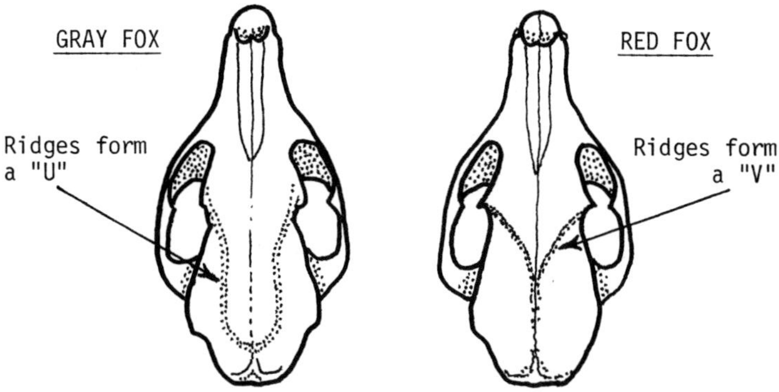

25 - a. 6 cheek teeth, 4th and 5th both larger than 6th; braincase narrows toward the back: --------- RACCOON
b. 5 cheek teeth, last 2 nearly equal in size, but larger than the first 3; braincase narrows toward the back of the skull: ---------------- 26
c. 4 cheek teeth, last 2 nearly equal in size, but larger than the first 2; braincase widest at back of skull, across occipital region: --- 27

26 - a. Length of rostrum about equal to diameter of orbit; occipital crest extends back on each side farther than end of sagittal crest; ear bones flattish: ----------------------- OTTER
b. Length of rostrum about 1½x the diameter of the orbit; sagittal crest extends back as far as, or farther than, the occipital crest; ear bones rounded, bulbous:
- Skull less than 96 mm long: --------------- MARTEN
-- Skull 96 to 125 mm long: ------------------ FISHER
--- Skull over 130 mm long: ---------------- WOLVERINE

27 - a. Ear bone rounded, bulbous; infraorbital foramen in front of orbit; first cheek tooth has 2 roots (or alveoli): ------------------ BADGER
b. Ear bone flattish; infraorbital foramen below front edge of orbit; first cheek tooth with 1 root; along Pacific coast: -------- SEA OTTER

28 - a. Only 1 incisor (just 2 across front of skull); palate complete between all cheek teeth; side of rostrum solid bone: Rodents ------------------ 29

b. One large and one very tiny incisor on each side (total of 4 across front of skull); palate only a short bridge between the 1st three cheek teeth; side of rostrum an open lattice-work of delicate bone: -------- HARE, or JACKRABBIT

- In some areas, could possibly be a DOMESTIC RABBIT.

JACKRABBIT

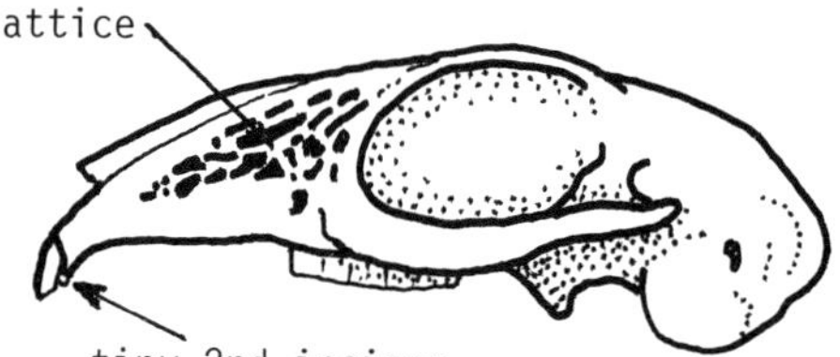

29 - a. Infraorbital foramen quite small, low on the rostrum and close to the palate: ------------------ 30

b. Infraorbital foramen very large, larger than the foramen magnum, nearly as large as the orbit: ----------------------------------- 31

30 - a. Skull over 90 mm (3½") long; postorbital process indicated by no more than a small bump: ------------------------------------- BEAVER

b. Skull less than 90 mm long; postorbital process prominent, sharply pointed: --------------- MARMOT

31 - a. Posterior edge of palate lies between last cheek teeth; palate very narrow between first cheek teeth, but present; paroccipitals short, normal: --------------------------------- PORCUPINE

b. Posterior edge of palate lies <u>behind</u> last cheek teeth; right and left first cheek teeth nearly touch at midline; paroccipital processes very large, over 10 mm long: ------------ NUTRIA

32 - a. Canines large, prominent; incisors small: --------------- 33

b. NO canines; 1-2 large incisors; broad toothless space between incisors and cheek teeth --- 36

c. One large incisor on each side, followed by a row of sharply pointed cheek teeth; NO broad toothless space between incisors and cheek teeth: ----------------------------------- MOLE

SKULLS

33 - a. 3 incisors, 3-6 cheek teeth: ---------------------------- 34
b. 5 incisors, 6-7 cheek teeth: ----------------------- OPOSSUM
c. 1 or 2 incisors on each side, separated from each other by a distinct notch in the premaxillas: --- BAT

34 - a. Palate extends posteriorly only about 4 mm beyond the last cheek tooth: ------------------------- 35
b. Palate extends posteriorly well beyond the last cheek tooth: ---
- Skull over 58 mm (2¼") long: ------------------- MINK
(In the central plains states, it might very rarely be a BLACK-FOOTED FERRET; elsewhere, it might sometimes be from a DOMESTIC FERRET. Both these species are very difficult to separate from the MINK)
-- Skull 40-58 mm long: ------------ LONG-TAILED WEASEL
--- Skull less than 40 mm (1½") long: --- ERMINE or LEAST WEASEL

35 - a. 6 cheek teeth; postorbital process distinct: ------ RINGTAIL
b. 4 cheek teeth; Skunks ---
- Skull less than 65 mm (2½") long; small, distinct postorbital process: ----------- SPOTTED SKUNK
-- Skull over 65 mm long; postorbital process only a slight bump; nasals straight when viewed in profile: ---------------------- STRIPED SKUNK
--- As above, but nasals concave in profile:HOG-NOSED SKUNK

36 - a. Only 1 large incisor tooth on each side (total of 2 incisors at front of mouth); side of rostrum solid bone; palate complete between cheek teeth: ----- 37
b. One large and another very tiny incisor just behind it on each side (total of 4 incisors at front of mouth); side of rostrum a lattice of delicate bone; palate a short bridge between the first 2-3 cheek teeth: Rabbits -------------------------- 46

37 - a. Infraorbital foramen a small hole or slit, low on the side of the rostrum, just in front of the base of the zygomatic arch: ------------------------- 38
b. IOF a roundish hole in the middle of the rostrum, extending through to the other side: ----------------- 42
c. IOF a slit-like opening high on the rostrum, between the posterior edge of the nasal bones and the zygomatic arch: ------------------------------------- 43
d. IOF a large hole, as large as or larger than the foramen magnum: ---
- IOF round: ------------------------------ CHINCHILLA
-- IOF 2x wider than it is high: ------------ GUINEA PIG

SKULLS

INFRAORBITAL FORAMEN

POCKET GOPHER

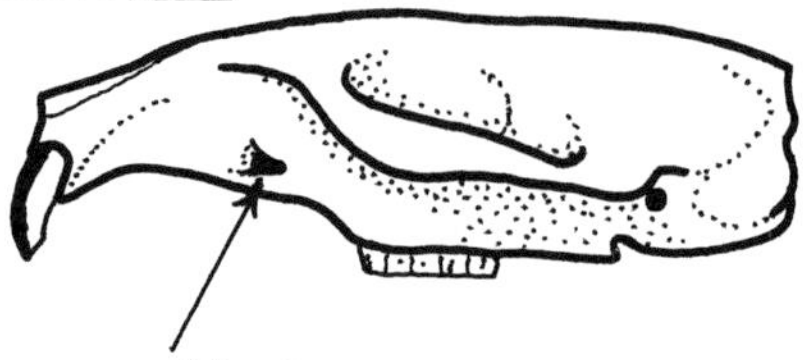

small, low on side of rostrum

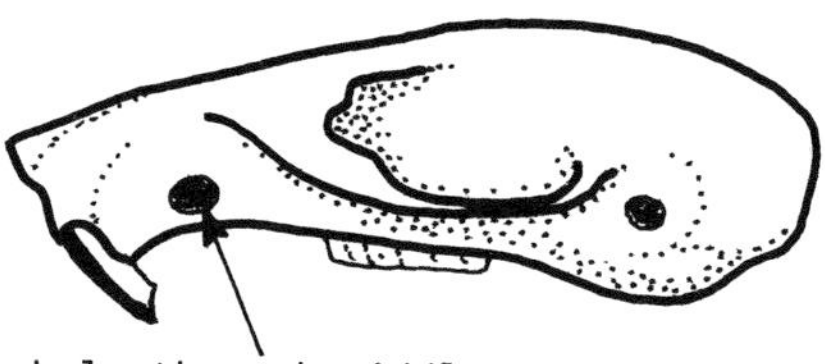

KANGAROO RAT

hole through middle of rostrum

slit-like, high on rostrum

MUSKRAT

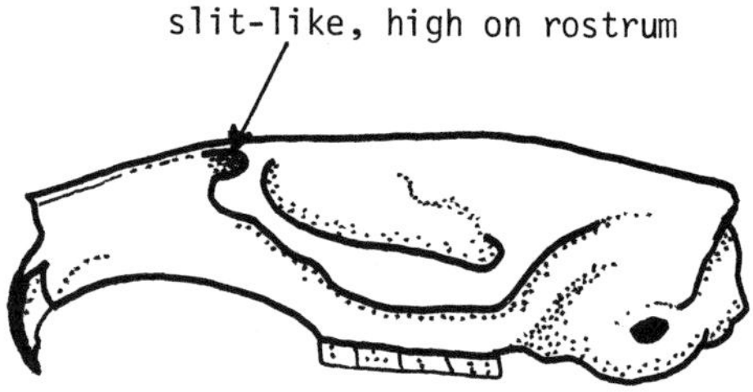

38 - a. Postorbital processes distinct, sharply pointed: Squirrels ------------------------------------ 39
b. NO postorbital processes, or only a low, rounded bump to indicate where they should be: ------- 41

39 - a. Cheek tooth rows essentially parallel; zygomatic arch vertical in cross section near its widest point: Tree Squirrels ---
- Skull over 60 mm long: ------ GRAY or FOX SQUIRREL
-- Skull 47 to 53 mm long: ----------- PINE SQUIRREL
--- Skull less than 45 mm long: ------ FLYING SQUIRREL
b. Cheek tooth rows closer together toward the rear; zygomatic arch horizontal in cross section near its widest point: ------------- 40

SKULLS

40 - a. Zygomatic width 40 mm or more: ----------------- PRAIRIE DOG
b. Zygomatic width less than 40 mm: Ground Squirrels
- Skull over 45 mm long: --------- Large GROUND SQUIRREL
-- Skull less than 45 mm long: ---- Small GROUND SQUIRREL
(This name includes also CHIPMUNKS, MANTLED GROUND SQUIRRELS, and the ANTELOPE GROUND SQUIRRELS, all of which have very similar skulls)

41 - a. Front surface of incisor is farther forward, or at least as far forward, as front tip of nasal bone; zygomatic arches nearly parallel, or perhaps slightly farther apart toward the front: ------------------- POCKET GOPHER
b. Nasal bone projects forward beyond the front surface of the incisor; zygomatic arches distinctly widest apart toward the rear: - MOUNTAIN BEAVER

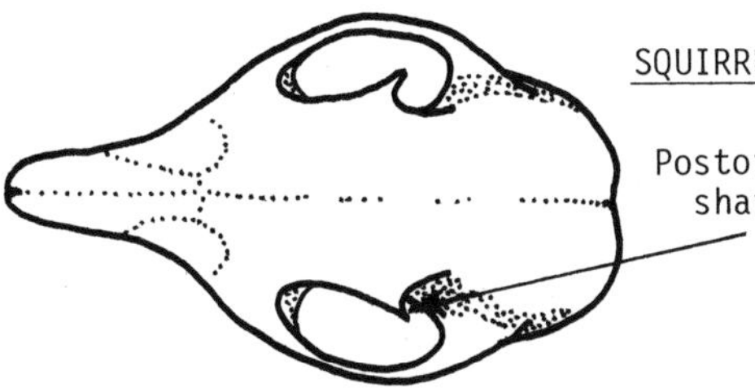

SQUIRREL

Postorbital process distinct, sharp, pointed.

MUSKRAT

Postorbital process low, rounded.

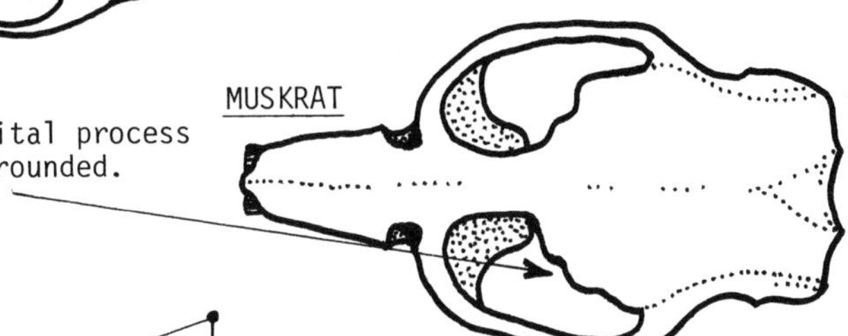

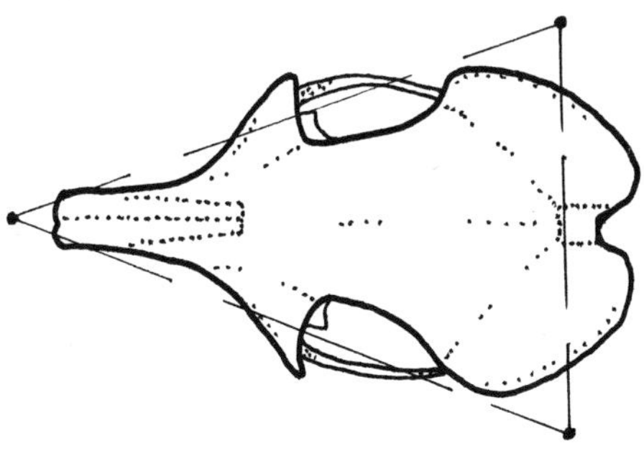

KANGAROO RAT

Inflated earbones, producing a roughly triangular shape.

SKULLS

42 - a. Earbones greatly inflated, so skull has a roughly triangular shape: ----------------- KANGAROO RAT
- If skull is less than 33 mm long, it is from a KANGAROO MOUSE.

b. Earbones NOT greatly inflated; skull NOT roughly triangular in shape: ---------------- POCKET MOUSE

43 - a. Postorbital processes small, at most only a small bump on front surface of the braincase: Microtine rodents ---
- Skull over 45 mm long: -------------------- MUSKRAT
-- Skull about 40 mm long: ------ ROUND-TAILED MUSKRAT
--- Skull less than 35 mm long: -- MEADOW MOUSE or VOLE (These two names also include about 15 species of voles and meadow mice)

b. NO sign of a postorbital process, so front of braincase flares smoothly back from interorbital area, without even a bump: ---------------------------- 44

44 - a. Skull less than 35 mm long: ------ WHITE-FOOTED or DEER MOUSE
- In arid regions, could also be the scarcer GRASSHOPPER MOUSE.
-- In wooded regions, and if each incisor has a lengthwise groove on the front surface, it could be a JUMPING MOUSE.

b. Skull over 35 mm long: ---------------------------------- 45

45 - a. Skull with a pair of ridges along the upper sides of the braincase, from interorbital region to occipital area; braincase is flattish between these ridges: Rats ---
- Skull over 38 mm long; nasal bones over 12 mm long: ------------------ HOUSE RAT
-- Skull <u>less</u> than 38 mm long; nasal bones <u>over</u> 12 mm long: ------------- RICE RAT
--- Skull less than 38 mm long; nasal bones less than 12 mm long: ------ COTTON RAT

b. NO ridges on braincase; skull over 35 mm long; nasals over 12 mm long: ------------------------ WOOD RAT

46 - a. Skull less than 45 mm long: ---------------------------- PIKA

b. Skull over 45 mm long: ------------------------------ RABBIT
- (Includes BRUSH RABBIT, COTTONTAIL, SWAMP RABBIT, MARSH RABBIT, and PYGMY RABBIT, which all have very similar skulls)
-- If the interparietal bone at the back of the braincase is fused with the parietals and occipital, rather than being visibly distinct, it is a SNOWSHOE HARE.

SKULLS

47 - a. A wide toothless space between the incisors
and the cheek teeth: Rodents ------------------------ 48
b. NO toothless space behind the incisors; all
teeth have sharp points: ------------------------------ 49

48 - a. Infraorbital foramen a rounded hole through
the middle of the rostrum: ------------------ POCKET MOUSE
b. Infraorbital foramen high on rostrum, at
base of zygomatic arch, near nasals: ---
- Incisors <u>NOT</u> grooved; a distinct
notch visible at working tip of
each incisor, when viewed in
profile; skull less than 23 mm
long: -------------------------------- HOUSE MOUSE
-- Incisors have a lengthwise groove
on their front surfaces; NO
notch at working tip of incisors;
skull less than 23 mm long: --------- HARVEST MOUSE
--- Incisors grooved as above; NO
notch at working tip; skull is
over 23 mm long: ------------------- JUMPING MOUSE

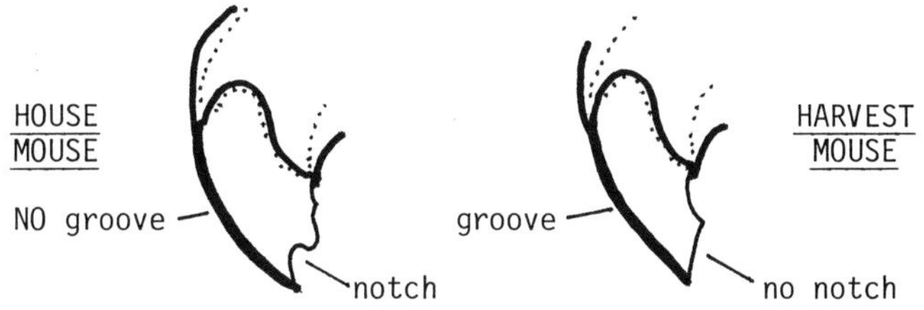

49 - a. Skull low, slender, without zygomatic arches;
only 2 teeth at front of jaw, the rest
forming two rows behind the first incisors: ---
- Teeth tipped with reddish-brown pigment: ---- SHREW
-- Teeth all white, not pigmented: -------- SHREW-MOLE
b. Skull chunky, zygomatic arches present;
incisors and premaxillas separated from each
other at front of skull by a distinct notch: --------- BAT

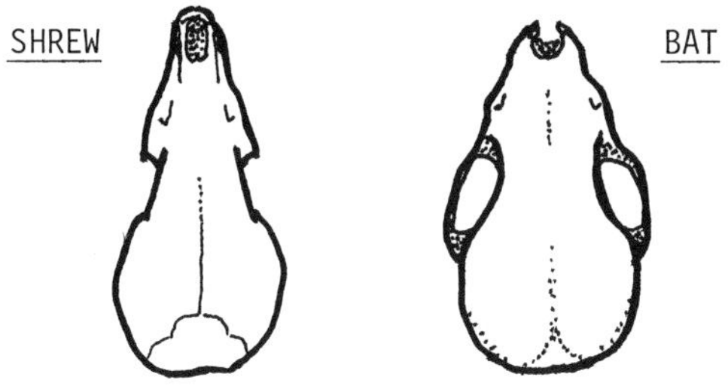

NO zygomatic arches   Notch in premaxillas

KEY-GUIDE to LOWER JAWS

The lower jaw, or mandible, is a fairly simple structure when compared to the skull. It consists of two dentary bones which meet at the front, joined at the mandibular symphysis. Each dentary consists of a horizontal ramus, which carries the lower teeth, a vertical portion called the coronoid process, a connecting point where the jaw joins the skull, called the condylar process, or more simply, the condyle, and a lower back corner referred to as the angle of the jaw.

Because there are only a few distinguishable parts to the lower jaw, there is less to use when attempting to identify the mammal from which it came. Consequently, identifications based on just the lower jaw are less certain than those based on the skull. By keying out both skull and lower jaw to the same species you are more likely to arrive at a correct identification.

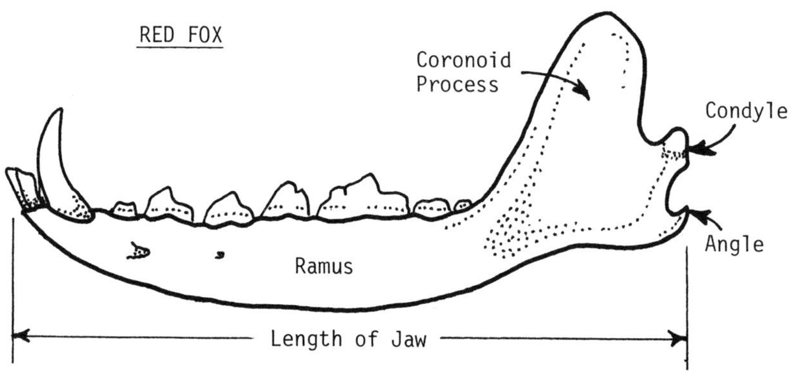

1 - a. Large jaws, over 100 mm (4") long: ---------------------- 2
    b. Medium jaws, 25-100 (1-4") long: ------------------------ 14
    c. Small jaws, less than 25 mm (1") long: ------------------ 33

2 - a. A long toothless space, more than 12 mm (½") long, in front of main row of cheek teeth (do not include 1-2 small peg-like teeth behind the canine as part of the main cheek tooth row): ---------------------------------- 3
    b. NO long toothless space in front of cheek teeth, or none more than 12 mm long: ----------------- 10

3 - a. Canine tooth large, prominent, and over 3x the height of the largest incisors: ---------------------- 4
    b. NO canine tooth, or if present it is about the same size as the incisors, or smaller: --------------- 5

LOWER JAWS

4 - a. Coronoid process high, rising 25 mm (1") or more above the condyle: ---------------------------------- BEAR

b. Coronoid process low, less than 25 mm higher than the condyle:

- Tip of coronoid process with a point, which is curved toward rear; condyle is most posterior part of jaw: --------------- PIG

-- Tip of coronoid process is blunt; back edge of angle is most posterior part of jaw: ---------------------------------- PECCARY

5 - a. 5-6 cheek teeth; 3 incisors; a canine may be present, shaped like another incisor: ------------------ 6

b. 4 flat-topped cheek teeth; 1 chisel-shaped incisor, colored on front surface; NO canine: ---

- Coronoid process higher than condyle; angle broad, rounded toward the rear: ----- BEAVER

-- Condyle higher than coronoid process; angle narrow, turned inward toward the rear: ----------------------------- PORCUPINE

6 - a. Coronoid process relatively straight, pointing upward; first and last cheek teeth about the same size: ---------------------------- HORSE or DONKEY

b. Coronoid process curves backward over condyle; first cheek tooth half the size of the last: ----------- 7

7 - a. Depth of ramus below the last cheek tooth nearly 2x the depth below the first cheek tooth; lower edge of angle higher than the lowest part of the ramus, so ramus is convex below: ----------- 8

b. Depth of ramus below first and last cheek teeth nearly equal; lower edge of angle on about the same level as lowest part of ramus, ramus thus even slightly concave: ---------------------------- 9

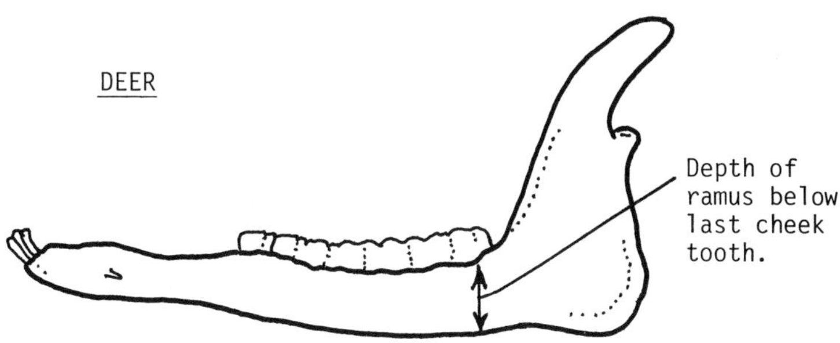

LOWER JAWS

8 - a. Jaw over 300 mm (1 ft) long: -------------------- COW or BISON
b. Jaw less than 300 mm long: ---
- Toothless space in front of cheek teeth less than 60 mm (2½") long: -- SHEEP or GOAT
-- Toothless space in front of cheek teeth more than 60 mm long: ------------ PRONGHORN

9 - a. Jaw less than 250 mm (10") long: ------------------------ DEER
b. Jaw 250-350 mm long: ------------------------------------- ELK
(in parts of Canada, could be CARIBOU)
c. Jaw more than 350 mm (14") long: ---------------------- MOOSE

10 - a. 3 incisors - Carnivores: ---------------------------------- 11
b. 2 incisors: --- 12
c. 1 incisor: -- 13

11 - a. 3 cheek teeth: --------------------------------- MOUNTAIN LION
b. 6-7 cheek teeth; angle is most posterior part of jaw: ---
- Jaw less than 115 mm (4½") long: ----------- RED FOX
-- Jaw more than 115 mm long: ------------------ COYOTE
(If lower edge of ramus is distinctly curved, convex, it could be from a DOG or WOLF; if the jaw is less than 175 mm long, it probably is from a DOG)
c. 6-7 cheek teeth; coronoid process, or both coronoid and angle together, are most posterior part of jaw: --- WOLVERINE

12 - a. Cheek teeth white, crowded together, each with 3-4 distinct points: ------------------------- HARBOR SEAL
b. Cheek teeth stained brownish, spaced apart, with only 1 main point (may be 1-2 tiny points at the gum line): ---
- Jaw over 220 mm long: ------------ NORTHERN SEA LION
-- Jaw less than 220 mm long; coronoid process is rounded: ---------- CALIFORNIA SEA LION
--- Jaw less than 220 mm long; coronoid with a blunt point at top: ----- NORTHERN FUR SEAL

SEA LION

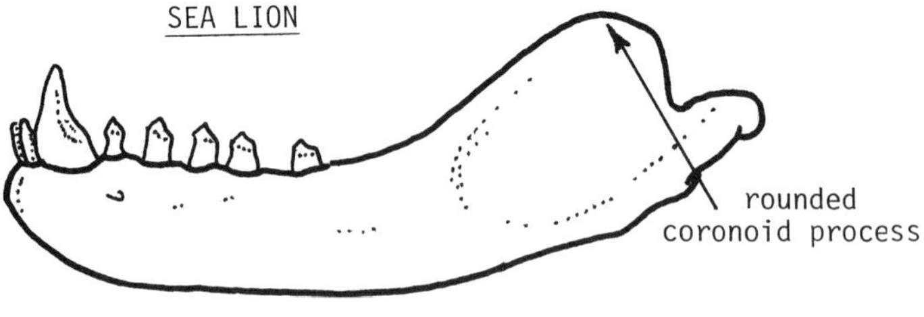

LOWER JAWS

13 - a. Coronoid process low, its apex well forward of
the condyle; Pacific coast: --------------- ELEPHANT SEAL

b. Coronoid process high, apex projecting back
over the condyle; Atlantic coast: ------------- GRAY SEAL

14 - a. Large, distinct canine; several small incisors: --------- 15

b. NO canine; 1 large chisel-shaped incisor;
a long toothless space behind incisor: ---------------- 21

c. Canine small, same size as incisors; all
cheek teeth with sharp points, set
close together: ------------------------------------ MOLE

d. NO canine, NO incisors; 7-8 peg-like cheek
teeth, spaced slightly apart: ----------------- ARMADILLO

15 - a. 3 cheek teeth: Cats ---
- Jaw less than 65 mm (2½") long: ------ DOMESTIC CAT
-- Jaw more than 65 mm long: ---------- LYNX or BOBCAT

b. 4-5 cheek teeth: -- 16

c. 6 cheek teeth: --- 19

d. 7 cheek teeth: --- 20

16 - a. Jaw over 60 mm long: ---------------------------------- 17

b. Jaw less than 60 mm long: ------------------------------ 18

17 - a. 3 incisors; angle and condyle project
toward the rear about equally: ---
- Height of coronoid process about
equal to width across its base;
back edge of coronoid process
has a slight break, or angle: ------------- BADGER
-- Height of coronoid process about
2x its width across the base;
back edge of coronoid process
follows a smooth curve: -------------------- OTTER

b. 2 incisors; condyle projects backward farther
than does the angle; Pacific coast: ----------- SEA OTTER

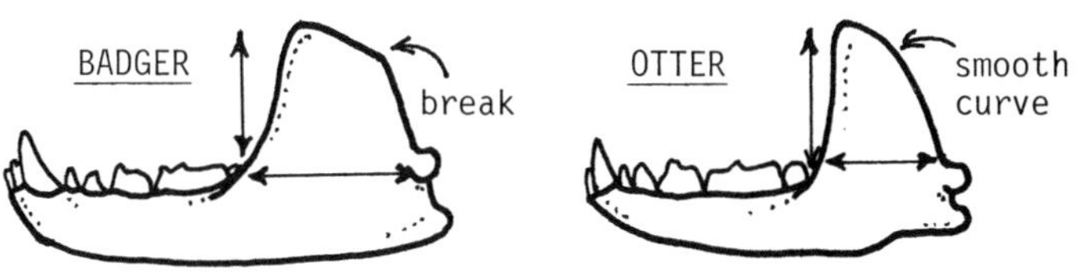

Note width and height of coronoid processes.

LOWER JAWS

18 - a. Back edge of coronoid process convex; entire coronoid process is roughly triangular: Weasels ---
- Jaw less than 35 mm long: ----------------- WEASEL
-- Jaw 35-50 mm long: ---------------- FERRET or MINK

b. Back edge of coronoid process straight, or slightly concave: Skunks ---
- Jaw less than 40 mm long: ----------- SPOTTED SKUNK
-- Jaw more than 40 mm long: ----------- STRIPED SKUNK

19 - a. Jaw less than 65 mm (2½") long: ---
- Back edge of coronoid process concave: --- RINGTAIL
-- Back edge of coronoid process convex: ------ MARTEN

b. Jaw more than 65 mm long: ---
- Jaw 65-90 mm long; coronoid process concave along back edge, apex above condyle: -------------------------------- RACCOON
-- Jaw about 80 mm long; apex of coronoid process distinctly in front of the condyle, and about 20 mm higher than the condyle: ----------------------- FISHER
--- Jaw over 90 mm long; back edge of coronoid process rises straight up from condyle, and apex is more than 25 mm higher than the upper edge of the ramus: ------------------- WOLVERINE
---- Jaw over 90 mm long; apex of coronoid process distinctly in front of condyle, rising about 10-15 mm above the condyle: ----------------------- COATI

20 - a. 3 incisors; angle is a hook-like process, at the end of a slight extension of the ramus:
- Lower edge of the ramus curves gently back to the end of the angle: ----------- RED FOX
(If the jaw is less than 90 mm long, it is probably from a SWIFT FOX or KIT FOX)
-- Lower edge of ramus with a distinct step, or break, just in front of the angle:----------------------------- GRAY FOX

b. 4 incisors; lower edge of ramus, including angle, turns sharply inward, toward midline: ----- OPOSSUM

LOWER JAWS

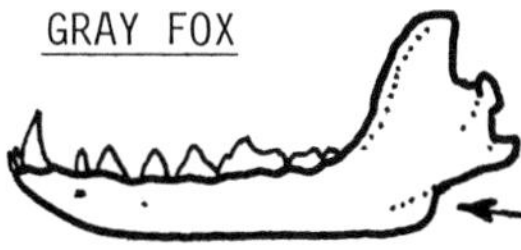

21 - a. NO distinct coronoid process - coronoid and condyle essentially a single structure: Rabbits ---
- Jaw over 50 mm (2") long: ------ HARE or JACKRABBIT
-- Jaw 35-50 mm long: ------------------------ RABBIT
(This name includes several species of cottontails and small rabbits)
--- Jaw less than 35 mm long: ------------------- PIKA

b. Coronoid process and condyle each distinct, separate structures: --------------------------------- 22

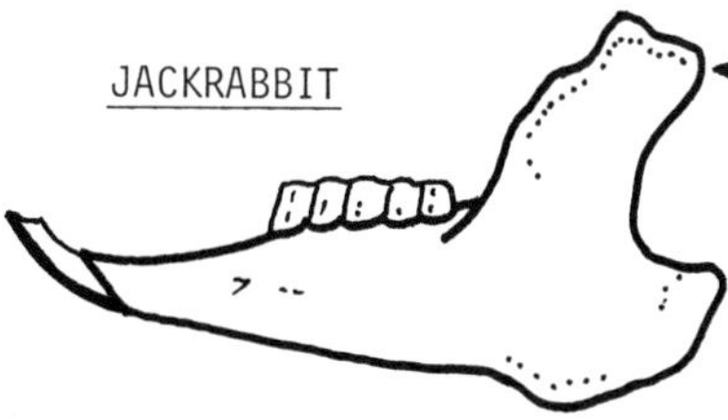

22 - a. Jaw over 38 mm (1½") long: ------------------------------ 23
b. Jaw 25-38 mm long: -- 30

23 - a. Condyle typical, a single distinct structure: ------------ 24
b. Condyle subdivided into 2 small bumps; angle a small, laterally projecting flange: ------ POCKET GOPHER

24 - a. Angle is clearly the most posterior part of the jaw: -- 25
b. Condyle is most posterior part of jaw, OR both condyle and angle project toward the rear about equally: ----------------------------------- 29

25 - a. Coronoid process higher than condyle, or both about equal in height: ------------------------- 26
b. Condyle distinctly higher than coronoid process: -- PORCUPINE

26 - a. A section of the lower edge of the ramus, just in front of the angle, curves inward; angle projects backwards, convexly: ------------------------ 27
b. Lower edge of ramus straight, or very tightly curled inward; angle points sharply upward: ----- MUSKRAT

LOWER JAWS

27 - a. Distinct bump on lower edge of ramus,
below first cheek teeth: ------------------------- BEAVER
b. NO bump along lower edge of ramus: ---------------------- 28

BEAVER

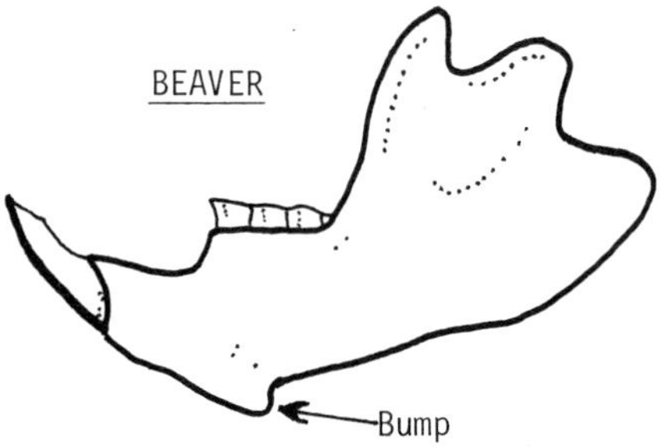

MUSKRAT

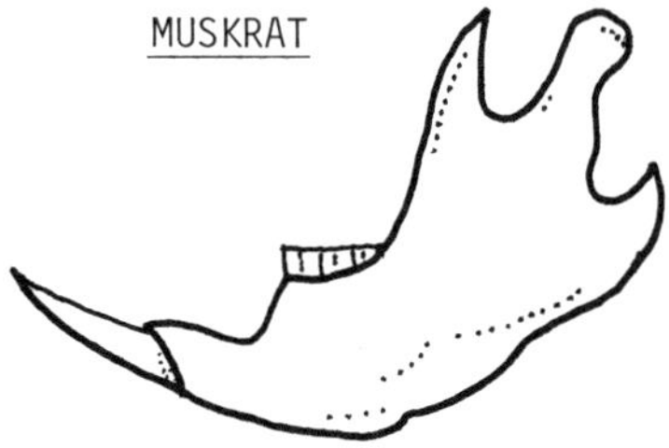

28 - a. Angle lies in same plane as rest of ramus: ---
- Jaw over 65 mm ($2\frac{1}{2}$") long: ------------------- MARMOT
-- Jaw less than 65 mm long: ---------------- PRAIRIE DOG
b. Angle turned outward, while lower edge
of ramus turns inward, both in the same
nearly horizontal plane: ----------------- MOUNTAIN BEAVER

29 - a. Coronoid process long; notch between the
coronoid process and the condyle more
than 4 mm deep: ------------------------- GROUND SQUIRRELS
b. Coronoid process short; notch between
coronoid process and condyle less
than 4 mm deep: --------------------------- TREE SQUIRRELS

30 - a. Angle lies at about the level of the cheek
teeth; lower edge of ramus is convex: ---------------- 31
b. Angle lies below the level of the cheek
teeth; lower edge of ramus straight
or slightly concave: ----------------------------------- 32

31 - a. Jaw less than 32 mm ($1\frac{1}{4}$") long: -------- MEADOW MOUSE or VOLE
b. Jaw more than 32 mm long: -------------- ROUND-TAILED MUSKRAT

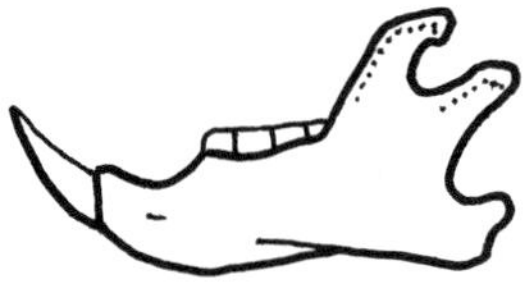

GROUND SQUIRREL

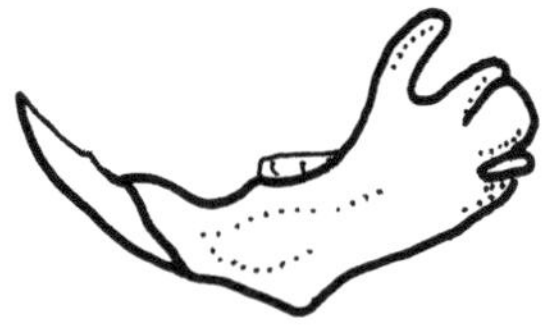

POCKET GOPHER

LOWER JAWS

32 - a. Section of lower edge of ramus, just in front of angle, curves or turns inward: ---
- Tip of coronoid process lies about 4-5 mm in front of angle: Small GROUND SQUIRRELS or CHIPMUNKS
-- Tip of coronoid process lies nearly above the angle: ---------------------------- PINE SQUIRREL
(If the jaw is less than 30 mm long, it is from a FLYING SQUIRREL)

b. Lower edge of ramus essentially straight, or with only a slight inward curl - NOT an entire section curled inward: ---
- Angle is most posterior part of jaw: --------- HOUSE RAT
-- Condyle is most posterior part of jaw, OR condyle and angle about equally far back: --- WOODRAT

33 - a. NO distinct canine; all teeth in front part of tooth row quite small, except the first teeth, which are larger: ---
- Teeth tipped with reddish-brown pigment: --------- SHREW
-- Teeth all white, NOT pigmented: ------------------- MOLE

b. A distinct canine present, larger than the incisors: ------ 34

c. NO canine, only 1 incisor; a toothless space between incisor and cheek teeth: --------------------- 35

34 - a. Condyle a short horizontal bar; coronoid process tall, roughly triangular: ------------------------ WEASEL

b. Condyle a small bump; coronoid process usually same height as condyle, or higher: ------------------ BAT

35 - a. Condyle subdivided into 2 small bumps; angle a short, laterally projecting flange: -------- POCKET GOPHER

b. Condyle NOT subdivided into bumps; angle does NOT project laterally: ------------------------------- 36

36 - a. A section of the lower edge of the ramus, just in front of the angle, turned inward: --- Small GROUND SQUIRRELS or CHIPMUNKS

b. Angle broad, flares outward and twists toward the horizontal plane: ---------- KANGAROO RAT or POCKET MOUSE

c. Angle in line with rest of jaw, NOT turned in, NOT flared out: --------------------------------- 37

37 - a. Condyle, or both condyle and angle, are most posterior part of jaw: ------------------------------- 38

b. Angle is clearly most posterior part of jaw: ---
- Jaw over 12 mm (½") long: -------------------- COTTON RAT
-- Jaw less than 12 mm long: ------------------- HOUSE MOUSE

LOWER JAWS

38 - a. Lower edge of ramus straight or slightly concave: --- 39

b. Lower edge of ramus below cheek teeth is clearly convex: --------------------- MEADOW MOUSE or VOLE

39 - a. Coronoid process very small, low: ---

- Jaw over 12 mm long: --- WHITE-FOOTED or DEER MOUSE

-- Jaw less than 12 mm long: --- HARVEST MOUSE or JUMPING MOUSE

b. Coronoid process larger, taller than condyle: ------------------------------- GRASSHOPPER MOUSE

SCIENTIFIC NAMES of the SPECIES in the KEY-GUIDES

The names given here are essentially those listed in the latest checklist of North American mammals, by Jones, et al (see References).

ANTELOPE GROUND SQUIRREL - *Ammospermophilus* (4 species)

ARMADILLO - *Dasypus novemcinctus*.

BADGER - *Taxidea taxus*.

BARBARY SHEEP - *Ammotragus lervia*. Also known as Aoudad.

BAT - There are about 40 species of bats known from the United States, but this key-guide is not designed to identify them to species; other sources are required (see References).

BEAVER - *Castor canadensis*.

BEAR - *Ursus americanus*, *U. arctos*, or *U. maritimus*.

BISON - *Bison bison*.

BLACK BEAR - *Ursus americanus*.

BOBCAT - *Felis rufus*. (formerly *Lynx rufus*)

CALIFORNIA SEA LION - *Zalophus californianus*.

CARIBOU - *Rangifer tarandus*.

CAT - Domestic cat, sometimes named *Felis catus*.

CHIPMUNK - *Tamias* (20 species in the western United States, plus 1 species in the east; all western species were formerly referred to as *Eutamias*)

COATI - *Nasua nasua*.

COTTON RAT - *Sigmodon* (4 species)

COTTONTAIL - *Sylvilagus* (4 species; see also RABBIT)

COW - Domestic cow or bull, sometimes named *Bos taurus*)

COYOTE - *Canis latrans*

DEER - *Odocoileus hemionus* (Mule Deer) or *O. virginianus* (White-tailed Deer). In some areas skulls or lower jaws which key out to DEER may be those of introduced AXIS DEER (*Cervus axis*), FALLOW DEER (*C. dama*), SIKA DEER (*C. nippon*), or others.

SCIENTIFIC NAMES

DEER MOUSE - *Peromyscus* (16 species). The name DEER MOUSE also includes here the GOLDEN MOUSE (*Ochrotomys nuttalli*) and the PYGMY MOUSE (*Baiomys taylori*)

DOG - Domestic dog, sometimes named *Canis familiaris*. Some breeds of dog will not key out in this key-guide, although they can usually be identified as carnivores, at least.

DOMESTIC CAT - see CAT, above.

DOMESTIC COW - see COW, above.

DOMESTIC DOG - see DOG, above.

DOMESTIC RABBIT - *Oryctolagus cuniculus*.

DONKEY - Domestic donkey, sometimes named *Equus asinus*.

ELEPHANT SEAL - *Mirounga angustirostris*.

ELK - *Cervus elaphus*. (formerly *C. canadensis*)

ERMINE - *Mustela erminea*.

FERRET - Domesticated ferret, usually named *Mustela furo*. Very rarely the skull of a BLACK-FOOTED FERRET (*M. nigripes*) might be encountered.

FISHER - *Martes pennanti*.

FLYING SQUIRREL - *Glaucomys volans* or *G. sabrinus*.

FOX SQUIRREL - *Sciurus niger*.

GOAT - Domestic goat, sometimes named *Capra hircus*.

GOPHER - see POCKET GOPHER, below.

GRASSHOPPER MOUSE - *Onychomys* (3 species)

GRAY FOX - *Urocyon cinereoargenteus*. On the islands off the coast of southern California may be found the smaller ISLAND FOX, *U. littoralis*.

GRAY SEAL - *Halichoerus grypus*.

GRAY SQUIRREL - *Sciurus carolinensis* in the eastern United States, *S. griseus* on the Pacific coast.

GRIZZLY BEAR - *Ursus arctos*. (formerly *U. horribilis*)

SCIENTIFIC NAMES

GROUND SQUIRREL - *Spermophilus* (22 species; in the past many of these species were given the name *Citellus*).

HARBOR SEAL - *Phoca vitulina*.

HARE - *Lepus* (8 species) Usually called JACKRABBIT; this group includes the SNOWSHOE RABBIT (*Lepus americanus*), which is really a hare, and the introduced EUROPEAN HARE (*L. capensis*).

HARVEST MOUSE - *Reithrodontomys* (5 species)

HOG-NOSED SKUNK - *Conepatus mesoleucus* or *C. leuconotus*.

HOODED SKUNK - *Mephitis macroura*

HORSE - Domestic horse, sometimes named *Equus caballus*.

HOUSE MOUSE - *Mus musculus*. The domesticated form is usually referred to as a LABORATORY MOUSE.

HOUSE RAT - *Rattus rattus* (ROOF or BLACK RAT) or *R. norvegicus* (NORWAY RAT). The domesticated LABORATORY RAT is a form of *R. norvegicus*.

JACKRABBIT - *Lepus* (8 species; see also HARE, above)

JUMPING MOUSE - *Zapus* (3 species) or *Napeozapus* insignis.

KANGAROO MOUSE - *Microdipodops megacephalus* or *M. pallidus*.

KANGAROO RAT - *Dipodomys* (15 species)

KIT FOX - *Vulpes macrotis*. Sometimes included with *V. velox*, the SWIFT FOX.

LEAST WEASEL - *Mustela nivalis*. (formerly *M. rixosa*)

LONG-TAILED WEASEL - *Mustela frenata*.

LYNX - *Felis lynx* (formerly *Lynx canadensis*)

MANTLED GROUND SQUIRREL - *Spermophilus lateralis* or *S. saturatus*. Sometimes called GOLDEN-MANTLED GROUND SQUIRREL; formerly named *Citellus lateralis* or *C. saturatus*.

MARMOT - *Marmota* (4 species) Includes the WOODCHUCK, *M. monax*.

MARTEN - *Martes americana*.

SCIENTIFIC NAMES

MEADOW MOUSE - *Microtus* (19 species); the term 'meadow mouse' also includes the SAGEBRUSH VOLE (*Lagurus curtatus*), BOG LEMMINGS (*Synaptomys*, 2 species), RED-BACKED VOLES (*Clethrionomys*, 3 species), HEATHER VOLE (*Phenacomys intermedius*), and the TREE VOLES (*Arborimus*, 2 species).

MINK - *Mustela vison*.

MOLE - *Scapanus* (3 species), *Parascalops breweri*, *Scalopus aquaticus*, or *Condylura cristata*.

MOOSE - *Alces alces* (formerly *A. americana*)

MOUNTAIN BEAVER - *Aplodontia rufa*.

MOUNTAIN GOAT - *Oreamnos americanus*.

MOUNTAIN LION - *Felis concolor*.

MULE DEER - *Odocoileus hemionus*. The BLACK-TAILED DEER is a Pacific coast form of the MULE DEER.

MUSKRAT - *Ondatra zibethicus*.

NORTHERN FUR SEAL - *Callorhinus ursinus*.

NORTHERN SEA LION - *Eumetopias jubatus*.

NUTRIA - *Myocastor coypus*.

OPOSSUM - *Didelphis virginiana*.

OTTER - *Lutra canadensis*.

PECCARY - *Tayassu tajacu*.

PIG - Domestic pig, sometimes named *Sus scrofa*. This name includes also feral or wild pigs.

PIKA - *Ochotona princeps*.

PINE SQUIRREL - *Tamiasciurus hudsonicus* or *T. douglasii*.

POCKET GOPHER - *Thomomys* (9 species in the western states), *Geomys* (5 species) or *Pappogeomys castanops* in the midwest to southeast.

POCKET MOUSE - *Perognathus* (18 species).

POLAR BEAR - *Ursus maritimus*.

PORCUPINE - *Erethizon dorsatum*.

SCIENTIFIC NAMES

PRAIRIE DOG - *Cynomys* (4 species).

PRONGHORN - *Antilocapra americana*.

RABBIT - *Sylvilagus* (8 species); includes COTTONTAILS, BRUSH RABBIT, MARSH RABBIT, SWAMP RABBIT, and PYGMY RABBIT.

RACCOON - *Procyon lotor*.

RED FOX - *Vulpes vulpes*. (formerly *V. fulva*).

RICE RAT - *Oryzomys* (3 species).

RINGTAIL - *Bassariscus astutus*.

ROUND-TAILED MUSKRAT - *Neofiber alleni*.

SEA OTTER - *Enhydra lutris*.

SHEEP - This term includes DOMESTIC SHEEP, sometimes named *Ovis aries*, BIGHORN SHEEP (*O. canadensis*), or DALL SHEEP (*O. dalli*). In BIGHORN males, the base of the horn or horn core is very large, larger than in domestic or DALL SHEEP. In female sheep, horn or horn cores are slender, and point more vertically. Many domestic breeds of sheep have no horns in either sex.
BARBARY SHEEP (*Ammotragus lervia*), introduced into several localities, have more laterally spreading horns, and they are less tightly curled.

SHREW - *Sorex* (24 species), *Blarina* (3 species), *Cryptotis parva*, or *Notiosorex crawfordi*.

SHREW-MOLE - *Neurotrichus gibbsii*.

SPOTTED SKUNK - *Spilogale gracilis* or *S. putorius*.

STRIPED SKUNK - *Mephitis mephitis*.

SWIFT FOX - *Vulpes velox*.

TREE SQUIRREL - *Sciurus* (6 species).

VOLE - This is the preferred common name for many species of MEADOW MOUSE and other closely related forms. See MEADOW MOUSE, above.

WEASEL - *Mustela erminea*, *M. nivalis*, or *M. frenata*.

WHITE-FOOTED MOUSE - *Peromyscus* (16 species). See also DEER MOUSE.

SCIENTIFIC NAMES

WHITE-TAILED DEER - *Odocoileus virginianus*.

WOLF - *Canis lupus* or *C. rufus*.

WOLVERINE - *Gulo gulo*. (formerly *G. luscus*).

WOODRAT - *Neotoma* (9 species).

CLEANING SKULLS

Clean, complete skulls and lower jaws are attractive and interesting. Some readers may wish to save the skulls from animals they find, especially if they are not broken and still have most of their teeth.

If a skull (or lower jaw) is already nearly stripped of flesh by insects when it is found, often a simple brushing, with an old toothbrush, soap, and running water, may be enough to clean it. Be sure all soft tissue is removed from the brain cavity and the nasal passages before allowing the scrubbed skull to dry.

If shreds of meat still cling to the skull, they should be removed. In most cases this is easily accomplished by cooking the skull. If it is a small skull, it can be dropped into boiling water for a few minutes; usually five minutes is enough. Then take the skull out of the water (with forceps or tongs), hold it under cool running water, and remove any bits of loose tissue with forceps, fingers, or whatever seems to work best. An old toothbrush is very useful here too.

Larger skulls should not be dropped into boiling water, since the sudden change in temperature may cause large teeth to split. They can be placed in cool water in a large pan or can, to which a bit of detergent has been added to cut grease. Place these on a stove or hot plate to warm up to a simmer -- no warmer. This should NOT be done in the kitchen -- better out in the garage, or outside. After several hours, or overnight, or even several days if the skull is very large, the soft tissue can be easily removed from the skull bones. A little trimming by hand under running water, and some brushing with a toothbrush or other small brush, will produce a nice clean skull.

Cleaned skulls can be left in the sun for a few days to dry and bleach a bit. They can also be soaked, for a few minutes only, in a solution of hydrogen peroxide (only a few drops in a gallon of water) to whiten them nicely. After the brief soaking, allow the skull to dry without rinsing. Commercial bleaches could also be used, but they often leave a white sediment or crust on the bone.

Skulls can also be cleaned by allowing them to rot in water, without any heating or cooking. This is usually a smelly process, not to be tried indoors, but the results are excellent. Be sure to allow sufficient time for all soft tissue to be rotted away.

Skulls can also be cleaned by allowing insects, particularly carrion beetles (Dermestids), or meal worms to feed on flesh which adheres to the bone. Again, this should be done only outdoors.

Avoid letting flies get to the soft tissue of a fresh skull. They may lay eggs on the flesh, which quickly develop into maggots, and the maggots will clean the skull of tissue. However, they also release materials which may stain or discolor the underlying bone.

CLEANING SKULLS

Properly cleaned skulls, accompanied by appropriate data on each, have a distinct scientific value for many kinds of studies. Data on each skull should include information about when it was found, where it was found, and the collector's name. In addition, as much other information as might be available should also be recorded, such as the animal's sex, how it died, what color it was (for species with a variable coat), and so on.

REFERENCES

Many books and publications about mammals have information about mammal skulls; some even have keys which may be helpful in identifying skulls. Many such keys, however, combine features of color, or fur with features of the skull; often both types of features are not available, so the key cannot be used effectively.

The following references are particularly useful for the identification of skulls from skull features alone:

Glass, Bryan P. Key to Skulls of North American Mammals. New edition. Oklahoma State University Bookstore. 53 pages.
(This is a fine skull key, but it requires more knowledge of technical terms used to refer to various skull parts.)

Hall, E. Raymond. The Mammals of North America. John Wiley and Sons, New York. 2nd edition, 2 volumes. 1181 pages, plus 180 pages of indexes. 1981.
(Contains keys to major groups, then to lesser groups, and ultimately to species; many of these keys require other information besides that available from the skull alone.)

Ingles, Lloyd G. Mammals of the Pacific States. Stanford University Press, Stanford. 506 pages. 1965.
(Contains an excellent key to skulls on pages 51-71, which works well in the region covered.)

For more information about the names and distribution of the mammals of the United States, the following are useful:

Burt, William H. and Richard P. Grossenheider. A Field Guide to the Mammals. Houghton Mifflin, Boston. 284 pages. 2nd edition. 1965.
(Good pictures of the animals, and poor photos of their skulls. The names and distribution maps are a bit out-of-date).

Jones, J.K., Jr., D.C. Carter, H.H. Genoways, R.S. Hoffman, and D. W. Rice. Revised Checklist of North American Mammals North of Mexico, 1982. Occasional Paper No. 80, The Museum, Texas Tech University. 22 pages.
(The source of the scientific names used in this booklet).